Simple Solutions.

Minutes a Day-Mastery for a Lifetime!

Level 3

Mathematics

2nd Edition

1st semester

Nancy L. McGraw

Bright Ideas Press, LLC
Cleveland, Ohio

Simple Solutions Level 3 Second Edition

1st semester

Printed in the United States of America

ISBN-13: 978-1-934210-14-7
ISBN-10: 1-934210-14-5

Cover Design: Dan Mazzola
Editor: Kimberly A. Dambrogio

Welcome to Simple Solutions

Note to the Student:

This workbook will give you the opportunity to practice skills you have learned in previous grades. By practicing these skills each day, you will gain confidence in your math ability.

Using this workbook will help you understand math concepts easier and for many of you, it will give you a more positive attitude toward math in general.

In order for this program to help you be successful, it is extremely important that you do a lesson every day. It is also important that you check your answers and ask your teacher for help with the problems you didn't understand or that you did incorrectly.

If you put forth the effort, Simple Solutions will change your opinion about math forever.

Lesson #1

1. $7 + 8 = ?$

2. $86 \bigcirc 93$

3.   Write the next number in the sequence. 10, 20, 30, …

4. $34 - 17 = ?$

5.   Draw a triangle. How many sides does it have?

6. $76 + 38 = ?$

7.   What fraction is shaded?

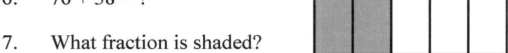

8.   Write the time shown on the clock.

9. $12 - 5 = ?$

10.   Sharon has 2 dimes and 1 nickel.
       How much money does she have?

11.   Would a watermelon weigh more than a
       pound or less than a pound?

12.   Write the standard number for $60 + 3$.

13.   Order these numbers from greatest to least.

             31   19   56   14

14.   Write the number that comes after 67.

15.   Monty has 2 fish bowls. There are 4 gold
       fish in each fish bowl. How many gold
       fish does Monty have altogether?

1.	2.	3.
4.	5.	6.
7.	8.	9.
10.	11.	12.
13.	14.	15.

Lesson #2

1. $364 - 193 = ?$

2. $6 + 8 = ?$

3.   Write the missing numbers in the sequence.

     5, ___, 15, 20, ____

4.   Write the standard number for $200 + 50 + 3$.

5. $7 - 4 = ?$

6.   Lisa had 12 flowers. She gave 4 to her teacher.
     How many flowers does Lisa have left?

7. $48 + 16 = ?$

8.   Name the shape.

9.   65 ◯ 37

10. $96 - 44 = ?$

11. Round 72 to the nearest ten.

12. What is the time shown on the clock?

13. Draw a square and shade $\dfrac{1}{4}$ of it.

14. Write 53 using words.

15. Jason has 2 quarters and 1 dime. How much money does he have?

1.	2.	3.
4.	5.	6.
7.	8.	9.
10.	11.	12.
13.	14.	15.

Lesson #3

1.   Round 87 to the nearest ten.

2. $455 + 361 = ?$

3.   Jessica had 55 red jelly beans and 48 green ones.
     How many jelly beans did she have in all?

4.   How many inches are in a foot?

5. $347 \bigcirc 374$

6. $83 - 27 = ?$

7.   What is the time 15 minutes after twelve noon?

8.   Which digit is in the tens place in 641?

9. $5 + 3 = ?$

10. Write the next number in the sequence. 3, 6, 9, ...

11. Twenty-five pennies are the same as a _____.

12. A triangle has _____ sides.

13. What fraction is shaded?

14. Is a paper clip about 1 inch
     long or 1 foot long?

15. Write 425 in expanded form.

1.	2.	3.
4.	5.	6.
7.	8.	9.
10.	11.	12.
13.	14.	15.

Lesson #4

1. $3 + 9 = ?$

2. What time is shown on the clock?

3. How many minutes are in 1 hour?

4. The answer to an addition problem is the _____.

5. $90 - 25 = ?$

6. Round 56 to the nearest ten.

7. $416 + 378 = ?$

8. Michael had 36 shells. Marisa gave him 19 more. How many shells does Michael have now?

9. 864 ◯ 846

10. $11 - 6 = ?$

11. What is the name of this shape?

12. Monique has 1 quarter, 1 dime, and 1 nickel. How much money does Monique have?

13. Write the number that comes before 314.

14. Which digit is in the thousands place in 7,439?

15. Put these numbers in order from least to greatest.

         28   97   54   63

1.

2.

3.

4.

5.

6.

7.

8.

9.

10.

11.

12.

13.

14.

15.

Lesson #5

1. $637 + 287 = ?$

2. Write the standard number for $600 + 90 + 3$.

3. $18 - 9 = ?$

4. What is the time 30 minutes after 1:00?

5. 74 ◯ 81

6. Round 534 to the nearest hundred.

7. $46 + 79 = ?$

8. Which digit is in the hundreds place in 8,352?

9. How many days are in a week?

10. Draw a rectangle and shade $\dfrac{3}{4}$ of it.

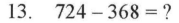

11. $\$27.98 - \$13.76 = ?$

12. Mark had 30 cupcakes. He gave 18 cupcakes to his classmates. How many cupcakes does he have left?

13. $724 - 368 = ?$

14. Would a bicycle weigh more than a pound or less than a pound?

15. Write 341 using words.

1.	2.	3.
4.	5.	6.
7.	8.	9.
10.	11.	12.
13.	14.	15.

Lesson #6

1.   Which digit is in the ones place in 9,163?

2. $8 + 4 = ?$

3. $81 - 37 = ?$

4.   Write the missing numbers in the sequence.

      4, 8, ____, 16, 20, ____

5.   Round 752 to the nearest hundred.

6. $314 + 583 = ?$

7.   Write the standard number for *seventy-six*.

8.   Brandon picked 68 strawberries and Marsha picked 97 strawberries. How many more strawberries did Marsha pick than Brandon?

9.   Mario has 3 nickels and 4 pennies. How much money does he have?

10. How many hours are in a day?

11. 198 ◯ 205

12. Look at this clock. Write the time.

13. Is 46 an even or an odd number?

14. Put these numbers in order from greatest to least.

           52   81   27   76

15. $235 - 97 = ?$

1.

2.

3.

4.

5.

6.

7.

8.

9.

10.

11.

12.

13.

14.

15.

Lesson #7

1. $9 - 3 = ?$

2. Round 463 to the nearest ten.

3. $28 + 56 = ?$

4. What time does the clock show?

5. Dante saw 17 boats in the harbor. Eight of the boats sailed away. How many boats were left?

6. What is the answer to an addition problem called?

7. $175 - 69 = ?$

8. Write 496 in expanded form.

9. $5 + 4 = ?$

10. Which digit is in the thousands place in 8,340?

11. Write the number that comes after 53.

12. Two nickels are equal to one _____ .

13. $74 + 9 = ?$

14. Is 53 an even number or an odd number?

15. Draw a circle. Shade $\frac{1}{2}$ of it.

1.	2.	3.
4.	5.	6.
7.	8.	9.
10.	11.	12.
13.	14.	15.

Lesson #8

1. Which is greater, 2 dimes or 1 quarter?

2. 739 $\bigcirc$ 793

3. Write the next number in the sequence. 6, 12, 18, _____

4. How many minutes are in a half-hour?

5. Six dimes and three pennies are how much money?

6. Would a giraffe's height best be measured in inches or in feet?

7. Which digit is in the tens place in 4,305?

8. $135 + 456 = ?$

9. On which day were the fewest tickets sold?

10. How many more tickets were sold on Thursday than on Wednesday?

11. $42 - 27 = ?$

Number of Circus Tickets Sold	
Mon.	🎟🎟🎟🎟🎟🎟
Tues.	🎟🎟🎟🎟
Weds.	🎟🎟🎟
Thurs.	🎟🎟🎟🎟🎟🎟🎟🎟
Fri.	🎟🎟🎟🎟🎟

Each ▯ = 5 tickets

12. **Two figures that have the same size and the same shape are called congruent.** Draw 2 congruent squares.

13. Round 5,836 to the nearest thousand.

14. Write 174 in expanded form.

15. Four nickels are equal to 2 _____.

1.	2.	3.
4.	5.	6.
7.	8.	9.
10.	11.	12.
13.	14.	15.

Lesson #9

1.   The answer to a subtraction problem is the _____.

2. $937 + 86 = ?$

3.   Round 4,364 to the nearest hundred.

4. $80 - 27 = ?$

5.   What number comes between 235 and 237?

6. $9 - 2 = ?$

7.   Which digit is in the thousands place in 37,215?

8.   Is 342 an even or an odd number?

9.   How long is the line segment?

10. Write the time 20 minutes after 3:00.

11. $63 + 58 = ?$

12. What fraction is shaded?

13. How many days are in a year?

14. $537 - 178 = ?$

15. Jackie bought 17 oranges, 16 apples and 9 peaches. How many pieces of fruit did she buy?

1.

2.

3.

4.

5.

6.

7.

8.

9.

10.

11.

12.

13.

14.

15.

Lesson #10

1. Is the number 865 even or odd?

2. $19 + 12 + 26 = ?$

3. How much time has passed from Clock 1 to Clock 2?

Clock 1   Clock 2

4. Round 7,826 to the nearest thousand.

5. $429 + 377 = ?$

6. What number comes before 816?

7. What fraction of the circle is shaded?

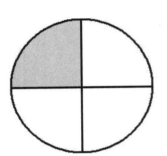

8. How many seconds are in a minute?

9. $3,432 + 6,599 = ?$

10. Which digit is in the ten thousands place in 26,514?

11. 563 ◯ 536

12. Donna has 3 quarters and 1 nickel. How much money does she have?

13. Write the standard number for $800 + 40 + 2$.

14. Seventeen children out of twenty-five children have brown hair. How many children do <u>not</u> have brown hair?

15. $107 - 58 = ?$

1.

2.

3.

4.

5.

6.

7.

8.

9.

10.

11.

12.

13.

14.

15.

Lesson #11

1. Look at this clock. Write the time.

2. 99 + 56 = ?

3. How many inches are in a foot?

4. Which digit is in the hundreds place in 36,507?

5. 8,366 + 4,597 = ?

6. Name the shape.

7. 10 – 6 = ?

8. Draw 2 congruent circles.

9. 512 – 264 = ?

10. What is the answer to an addition problem called?

11. Put these numbers in order from greatest to least.

    74   86   29   54   23

12. Jesse played 4 baseball games each week for 5 weeks. What is the total number of baseball games that Jesse played?

13. Is the number 37 even or odd?

14. Round 26,475 to the nearest thousand.

15. How much money is shown here?

1.	2.	3.
4.	5.	6.
7.	8.	9.
10.	11.	12.
13.	14.	15.

Lesson #12

1.   **There are 3 feet in 1 yard.** Write *3 feet = 1 yard* three times.

2. $35 + 21 + 14 = ?$

3.   There are 7 birds in the maple tree. There are twice as many birds in the oak tree. How many birds are in the oak tree?

4.   What is the time 25 minutes after 2:00?

5.   3,219 ◯ 3,192

6.   Write 732 using words.

7.   Round 5,447 to the nearest hundred.

8.   Would a desktop computer best be weighed in ounces or in pounds?

9. $17.25 − $10.54 = ?

10.   The answer to a subtraction problem is the _____.

11. $5,307 + 3,996 = ?$

12.   Which digit is in the hundred thousands place in 567,032?

13. $95 − 28 = ?$

14.   Write the next <u>two</u> numbers in the sequence. 18, 21, 24, …

15. $7 + 4 = ?$

1.	2.	3.
4.	5.	6.
7.	8.	9.
10.	11.	12.
13.	14.	15.

Lesson #13

1. Round 38,237 to the nearest ten thousand.

2. What is the name of this shape?

3. $3,775 + 2,819 = ?$

4. $26 - 19 = ?$

5. Write 479 in expanded form.

6. How many feet are in a yard?

7. How much time has passed from Clock 1 to Clock 2?

Clock 1   Clock 2

8. **All four-sided closed figures are called quadrilaterals.** Draw a closed shape that has four sides. Write *quadrilateral* under the shape.

9. Write the number that follows 7,348.

10. $12 - 3 = ?$

11. Which digit is in the ten thousands place in 90,378?

12. Write the standard number for $7,000 + 500 + 30 + 7$.

13. The difference is the answer to a(n) _____ problem.

14. $727 - 359 = ?$

15. What is the date on the third Thursday in February?

February

Sun	Mon	Tue	Wed	Thu	Fri	Sat
						1
2	3	4	5	6	7	8
9	10	11	12	13	14	15
16	17	18	19	20	21	22
23	24	25	26	27	28	

1.	2.	3.
4.	5.	6.
7.	8.	9.
10.	11.	12.
13.	14.	15.

Lesson #14

1. Which of the following is equal to 815?

   a) (800 + 50 + 1)   b) (800 + 10 + 5)

2. There were *eight thousand, three hundred seventy-seven* people at the concert. Write this number in standard form.

3. 6,547 + 2,863 = ?

4. A quadrilateral has _____ sides.

5. 33 – 7 = ?

6. Write the time.

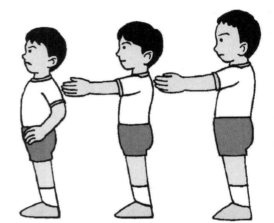

7. 6 + 9 = ?

8. Round 53,284 to the nearest hundred.

9. Would a feather best be weighed in ounces or in pounds?

10. How many days are in a year?

11. 47 + 58 = ?

12. Which digit is in the hundred thousands place in 703,216?

13. What fraction is not shaded?

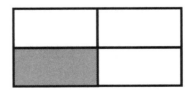

14. 8,364 ◯ 8,634

15. The boys lined up from tallest to shortest. Paul is taller than Perry. Hank is shorter than Perry. Greg is taller than Paul. In what order were the boys lined up?

1.	2.	3.
4.	5.	6.
7.	8.	9.
10.	11.	12.
13.	14.	15.

Lesson #15

1. Round $2.58 to the nearest dollar.

2. Write 4,216 using words.

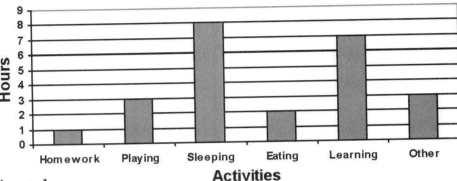

Student Daily Activities

4. How many more hours are spent sleeping than eating?

3. $93 + 37 = ?$

5. Which two activities take the same amount of time?

6. Write $70,000 + 6,000 + 300 + 20 + 1$ as a standard number.

7. $7,825 + 4,997 = ?$

8. How many feet are in a yard?

9. What will be the time 5 minutes before noon?

10. Four quarters are equal to one _____.

11. $607 - 326 = ?$

12. Write the name for a four-sided shape.

13. How many minutes are in an hour?

14. Draw a rectangle and shade $\frac{2}{5}$ of it.

15. Write the next number in the sequence. 10, 15, 20, …

1.

2.

3.

4.

5.

6.

7.

8.

9.

10.

11.

12.

13.

14.

15.

Lesson #16

1. Write 4,353 in expanded form.

2. $11 - 2 = ?$

3. If Ryan saves $0.25 each day for 5 days, how much money will he have at the end of the five days?

4. $198 + 378 = ?$

5. Is a door about 8 inches tall or 8 feet tall?

6. $7 + 7 = ?$

7. Which is greater, (6 tens and 3 ones) or 71?

8. $703 - 246 = ?$

9. Which digit is in the thousands place in 86,931?

10. A quadrilateral has _____ sides.

11. Round 375,988 to the nearest thousand.

12. Put these numbers in order from least to greatest.

    8,643   7,925   8,456   7,295

13. $5,036 + 8,879 = ?$

14. How many days are in a week?

15. What is the time?

1.	2.	3.
4.	5.	6.
7.	8.	9.
10.	11.	12.
13.	14.	15.

Lesson #17

1. The Smith family is eighth in a line of 12 families at an amusement park. Are there more families in front of them or behind them?

2. Write 675 using words.

3. $89 + 34 + 22 = ?$

4. $13 - 7 = ?$

5. Round 86,207 to the nearest ten thousand.

6. How many feet are in a yard?

7. $\$65.99 - \$27.37 = ?$

8. Write the time 15 minutes before 5:00.

9. Write $70,000 + 3,000 + 500$ in standard form.

10. $6,185 + 7,236 = ?$

11. 6,439 ◯ 6,394

12. A four-sided shape is called a(n) _____.

13. **There are 16 ounces in a pound.** Write *16 ounces = 1 pound* three times.

14. Which digit is in the ten thousands place in 169,347?

15. Draw 2 congruent squares.

1.

2.

3.

4.

5.

6.

7.

8.

9.

10.

11.

12.

13.

14.

15.

Lesson #18

1. How many ounces are in a pound?

2. $418 - 89 = ?$

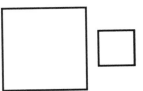

3. **Figures with the same shape, but different sizes are called similar.** Draw 2 similar squares.

4. Chris has 2 dollars, 2 quarters, and 3 dimes. How much money does he have?

5. Leon and Tyrone have 14 mice. Leon has 2 more mice than Tyrone. How many mice does each boy have?

6. $6 - 2 = ?$

7. Round 365,214 to the nearest hundred thousand.

8. $377 + 582 = ?$

9. The answer to a subtraction problem is the _____.

10. 36,517 ◯ 36,175

11. Write the first four even numbers.

12. Which digit is in the hundred thousands place in 451,368?

13. Write the standard number for $900,000 + 40,000 + 6,000 + 500 + 20 + 7$.

14. How many inches are in a foot?

15. $6,447 + 5,863 = ?$

1.

2.

3.

4.

5.

6.

7.

8.

9.

10.

11.

12.

13.

14.

15.

Lesson #19

1. Write the time 30 minutes after 7:00.

2. Draw 2 similar rectangles.

3. $7,716 + 8,947 = ?$

4. $14 - 6 = ?$

5. How many ounces are in a pound?

6. Round $9.74 to the nearest dollar.

7. $800 - 256 = ?$

8. Which digit is in the thousands place in 31,246?

9. The movie began at 8:30 p.m. It lasted one hour. At what time did the movie end?

10. A quadrilateral has _____ sides.

11. Rosie saved $5.25 the first week and $6.85 the second week. How much money did she save?

12. $5 + 4 = ?$

13. Would the length of a car be about 7 inches or 7 feet?

14. What is the name of this shape?

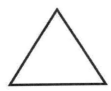

15. Put these numbers in order from least to greatest.

     3,416   3,641   3,146

1.

2.

3.

4.

5.

6.

7.

8.

9.

10.

11.

12.

13.

14.

15.

Lesson #20

1. $12 - 9 = ?$

2. If Jeff has 6 quarters, how much money does he have?

3. Write $8,000 + 400 + 10 + 9$ as a standard number.

4. Julie estimates that she saw 40 ducks on the pond. What is the smallest number of ducks she may have seen?

5. How many cookies are in a dozen?

6. $86,245 \bigcirc 86,452$

7. Round 57,313 to the nearest hundred.

8. $9,364 + 8,775 = ?$

9. Write the time.

10. $682 - 255 = ?$

11. Would a pencil best be weighed in ounces or in pounds?

12. Is 68 an even or an odd number?

13. Draw 2 congruent ellipses.

14. It is 3:00 now. What time will it be in 45 minutes?

15. $12 + 36 + 15 = ?$

1.

2.

3.

4.

5.

6.

7.

8.

9.

10.

11.

12.

13.

14.

15.

Lesson #21

1. $4 + 7 = ?$

2.   Carina went for a walk at 4:30. She got home at 5:10. How long was her walk?

3. $400 - 266 = ?$

4.   Round 347,251 to the nearest ten thousand.

5.   Write $50,000 + 4,000 + 900 + 60 + 7$ as a standard number.

6.   Joseph delivers newspapers to 213 homes. So far, he has delivered newspapers to 169 homes. To how many more homes does he still need to deliver newspapers?

7. $8,893 + 7,565 = ?$

8.   Would a horse best be weighed in ounces or in pounds?

9.   How many feet are in a yard?

10. $47 - 29 = ?$

11.   Two figures with the same size and shape are called _____.

12.   Which digit is in the hundreds place in 74,305?

13.   **5 + 5 + 5 means the same as 3 × 5.** They are both equal to what number?

14.   5,831 $\bigcirc$ 5,381

15.   What fraction is shaded?

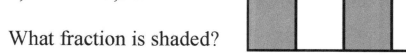

1.	2.	3.
4.	5.	6.
7.	8.	9.
10.	11.	12.
13.	14.	15.

Lesson #22

1. Draw 2 rhombuses that are similar.

2. $2 \times 7 = ?$

3. $6 - 2 = ?$

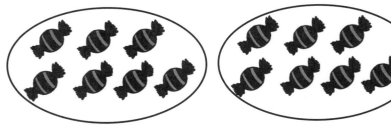

4. Write *seven thousand, six hundred eighty-two* as a standard number.

5. Leah has 1 quarter, 1 dime, 1 nickel, and 3 pennies. How much money does she have?

6. $7,758 + 3,892 = ?$

7. $5 + 8 = ?$

8. Round 53,937 to the nearest ten.

9. $841 - 373 = ?$

10. Write the time.

11. Write the next two numbers in the sequence. 15, 19, 23, …

12. Sixty-three children chose red as their favorite color. Forty-seven children chose blue. How many more children chose red than blue?

13. Write the number that comes before 860.

14. Write 4,215 in expanded form.

15. Would the length of a pencil be about 7 inches or 7 feet?

1.	2.	3.
4.	5.	6.
7.	8.	9.
10.	11.	12.
13.	14.	15.

Lesson #23

1. If it is 2:10 now, what time will it be in 20 minutes?

2. $5 \times 6 = ?$ (Hint: Think 5 groups of 6.)

3. How many ounces are in a pound?

4. $1,316 + 7,559 = ?$

5. In a bike race, Angela is behind Vince. Mike is ahead of Charles. Mike is between Angela and Charles. What is the order of the children in the race?

6. $805 - 364 = ?$

7. Any number multiplied by zero is equal to _____.

8. Round 9,317 to the nearest thousand.

9. Write a multiplication sentence for $2 + 2 + 2 + 2$ and solve it.

10. How many minutes are in an hour?

11. A quadrilateral has _____ sides.

12. What is the name of the shape?

13. **There are 4 quarts in a gallon.** Write *4 quarts = 1 gallon* three times.

14. How many dimes are in one dollar?

15. Write 3,254 using words.

1.	2.	3.
4.	5.	6.
7.	8.	9.
10.	11.	12.
13.	14.	15.

Lesson #24

1.   5,214 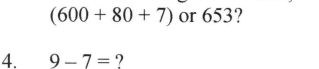 2,514

2.   Write the time shown on the clock.

3.   Which has the greater value,
     $(600 + 80 + 7)$ or 653?

4. $9 - 7 = ?$

5.   Write a multiplication sentence for $5 + 5 + 5 + 5 + 5$ and solve it.

6.   What is the measure of the line segment below?

7.   Draw 2 congruent hearts.

8.   Write the other members of the
     fact family for $5 + 7 = 12$.

9. $7,793 + 6,085 = ?$

10. $9 \times 1 = ?$ (Hint: Think 9 groups of 1.)

11. $15.69 - 9.99 = ?$

12. Round 937 to the nearest ten.

13. Would eating a healthy breakfast take
     about 15 seconds or 15 minutes?

14. What is the value of 5 quarters?

15. Lily has to pay 10 cents a day for each overdue library book. She
     has 3 books to return. They are one day late. How much money
     will Lily have to pay in fines?

1.

2.

3.

4.

5.

6.

7.

8.

9.

10.

11.

12.

13.

14.

15.

Lesson #25

1. $8 \times 0 = ?$

2.   What number follows 7,437?

3. $8 + 9 = ?$

4.   Write an addition sentence for 2×5 and solve it.

5.   How many quarts are in a gallon?

6.   How much money is shown?

7. $4,135 + 2,698 = ?$

8.   Antonio left home at 8:15. He arrived at school at 8:45. How long did it take him to get to school?

9.   How many days are in 2 weeks?

10. Put these numbers in order from greatest to least.

     865   438   917   377

11. $75 - 19 = ?$

12. Would a badminton birdie best be weighed in ounces or in pounds?

13. Round 32,186 to the nearest ten thousand.

14. Write $20,000 + 2,000 + 80 + 8$ as a standard number.

15. Marcie bought a notebook that cost $1.79. She gave the clerk $5. How much change should Marcie receive?

1.	2.	3.
4.	5.	6.
7.	8.	9.
10.	11.	12.
13.	14.	15.

Lesson #26

1. How much change will you get from $5.00 if you buy a magazine that costs $2.95?

2. How many months are in a year?

3. Does it take about 1 second or 1 minute to brush your teeth?

4. $2 \times 9 = ?$

5. Is 358 an even or an odd number?

6. How many nickels are in a dollar?

7. $8 + 7 = ?$

8. What time will it be at 5 minutes before noon?

9. $3,317 + 4,588 = ?$

10. Write 26,489 in expanded form.

11. Write a multiplication sentence for $4 + 4 + 4$ and solve it.

12. $605 - 371 = ?$

13. Which digit is in the thousands place in 718,254?

14. $10 + 31 + 18 = ?$

15. $65,386 \bigcirc 56,863$

1.	2.	3.
4.	5.	6.
7.	8.	9.
10.	11.	12.
13.	14.	15.

Lesson #27

1. $713 + 557 = ?$

2.   Should a flight from Cleveland, Ohio to Orlando, Florida take about 2 hours or 2 days?

3.   Monica began her piano lesson at 3:45 p.m. If her lesson lasts 30 minutes. At what time will her lesson be over?

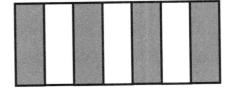

4.   Round $6.54 to the nearest dollar.

5. $3 \times 7 = ?$ (Hint: Think 3 groups of 7.)

6.   The answer to an addition problem is the _____.

7. $66 - 48 = ?$

8. $7 + 4 + 2 = ?$

9.   Is 2,651 an even or an odd number?

10. Marlon has 3 dollars, 2 quarters, and 2 dimes. How much money does Marlon have?

11. What fraction is shaded?

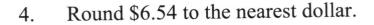

12. $6 \times 0 = ?$

13. Write $300,000 + 50,000 + 4,000 + 800 + 90 + 3$ as a standard number.

14. $5,632 + 6,775 = ?$

15. This clock shows what time?

1.	2.	3.
4.	5.	6.
7.	8.	9.
10.	11.	12.
13.	14.	15.

Lesson #28

1. List the first 4 odd numbers.

2. The answer to a subtraction problem is called the _____.

3. What time was it 15 minutes ago, if it is 12:30 now?

4. $28 + 96 = ?$

5. How many quarts are in a gallon?

6. $5 \times 2 = ?$ (Hint: Think 5 groups of 2.)

7. $13 - 6 = ?$

8. $2,883 + 977 = ?$

9. The westbound train carries 475 passengers. The eastbound train carries only 295 passengers. How many more passengers can ride the westbound train than the eastbound train?

10. Write a multiplication sentence for $3 + 3 + 3 + 3 + 3 + 3$ and find the solution.

11. Two figures having the same shape, but different sizes are _____.

12. **A line is a straight path that goes on without end in two directions.**   Draw a line.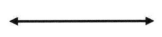

13. $935 - 588 = ?$

14. Draw a heart. Draw a line of symmetry on the heart.

15. How many inches are in 2 feet?

1.

2.

3.

4.

5.

6.

7.

8.

9.

10.

11.

12.

13.

14.

15.

Lesson #29

1. $601 - 388 = ?$

2. Which digit is in the ten thousands place in 471,235?

3. How many feet are in a yard?

4. $9,635 + 7,544 = ?$

5. Round 42,366 to the nearest hundred.

6. What time will it be in 15 minutes, if it is 1:15 now?

7. $79 + 38 = ?$

8. $4 \times 4 = ?$

9. Is it more likely that Marco lives 2 feet from school or 2 miles from school?

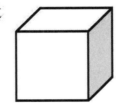

10. Look at this shape. What is its name?

11. Dina has 3 rosebushes. If she picks 2 roses from each bush, how many roses does she pick?

12. Draw a circle. Shade $\dfrac{1}{4}$ of it.

13. Is 3,507 an even or an odd number?

14. How many quarters are in $2?

15. $5 \times 0 = ?$

1.	2.	3.
4.	5.	6.
7.	8.	9.
10.	11.	12.
13.	14.	15.

Lesson #30

1. List the first four even numbers.

2. $3 \times 9 = ?$

3. Write the next number in the sequence. 25, 27, 29, …

4. $512 - 377 = ?$

5. Draw 2 similar rectangles.

6. $6 - 1 = ?$

7. Write 36,245 in expanded form.

8. Round 7,788 to the nearest ten.

9. Write the time.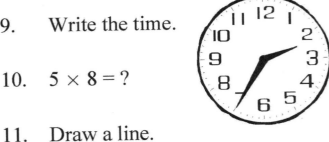

10. $5 \times 8 = ?$

11. Draw a line.

12. Between 2006 and 2007, Grady Sizemore had 1,283 at-bats. Victor Martinez had 1,134 at-bats in the same time frame. How many more at-bats did Grady Sizemore have than Victor Martinez?

13. How many ounces are in a pound?

14. $3,227 + 5,388 = ?$

15. What is the answer to a multiplication problem called?

1.	2.	3.
4.	5.	6.
7.	8.	9.
10.	11.	12.
13.	14.	15.

Lesson #31

1. $4 \times 6 = ?$

2. It is 5:10 now. What time was it 10 minutes ago?

3. $9 + 9 = ?$

4. $8,977 \bigcirc 8,797$

5. Write the number that comes before 590.

6. Beth spent $4.65 at the toy store. She gave the clerk $10. How much change should she get back?

7. $11 - 5 = ?$

8. Round 37,819 to the nearest ten thousand.

9. $700 - 457 = ?$

10. Put these numbers in order from least to greatest.

    886   688   686   866

11. What fraction of this figure is shaded?

12. A four-sided shape is called a(n) _____.

13. $4,076 + 4,281 = ?$

14. $3 \times 1 = ?$

15. Write the rest of the fact family for $5 + 6 = 11$.

1.	2.	3.
4.	5.	6.
7.	8.	9.
10.	11.	12.
13.	14.	15.

Lesson #32

1. Is 7,526 an even or an odd number?

2. $5 \times 9 = ?$

3. Which digit is in the tens place in 42,196?

4. Draw a line.

5. How many quarts are in a gallon?

6. $418 - 166 = ?$

7. $55 + 24 + 16 = ?$

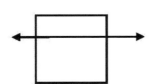

8. Write $40,000 + 6,000 + 500$ as a standard number.

9. 8,743 ◯ 8,473

10. Mr. Thomas had a board that was 36 inches long. He cut off an 8-inch piece. How long was the board after he made the cut?

11. The answer to an addition problem is called the _____.

12. $645 + 392 = ?$

13. At 15 minutes before 11:00, what is the time?

14. Does this diagram show a line of symmetry?

15. Round 313,866 to the nearest hundred thousand.

1.

2.

3.

4.

5.

6.

7.

8.

9.

10.

11.

12.

13.

14.

15.

Lesson #33

1. What is the answer to a multiplication problem called?

2. $6 + 9 = ?$

3. Write the time.

4. $56 + 39 = ?$

5. Write a multiplication sentence for $3 + 3 + 3 + 3 + 3 + 3$.

6. Marie has 3 pages of stickers with 5 stickers on each page. How many stickers does Marie have?

7. Draw a rectangle. Show one line of symmetry on the rectangle.

8. $723 - 588 = ?$

9. How many feet are in 2 yards?

10. Is it more likely that the bus ride to school takes 20 seconds or 20 minutes?

11. Write 329,265 in expanded form.

12. $5 \times 5 = ?$

13. Would a slice of bread best be weighed in ounces or in pounds?

14. $7,325 + 6,414 = ?$

15. Daniel has 4 dollars, 1 quarter, 3 dimes, and 3 nickels. How much money does he have?

1.	2.	3.
4.	5.	6.
7.	8.	9.
10.	11.	12.
13.	14.	15.

Lesson #34

1. $6,112 + 2,897 = ?$

2. $3 \times 8 = ?$

3. Draw a line.

4. David has 61 football cards and 29 baseball cards. How many more football cards than baseball cards does David have?

5. Two figures with the same size and shape are _____.

6. $386 - 179 = ?$

7. Any number multiplied by zero is equal to _____.

8. How many pennies are worth the same as a dollar?

9. What time will it be in 40 minutes, if it is 5:20 now?

10. $5 \times 2 = ?$

11. How many quarts are in 2 gallons?

12. Write 577 using words.

13. Is the number 36,248 even or odd?

14. Would a basketball hoop be about 9 inches or 9 feet high?

15. Round 37,842 to the nearest thousand.

1.	2.	3.
4.	5.	6.
7.	8.	9.
10.	11.	12.
13.	14.	15.

Lesson #35

1. What fraction is shaded?

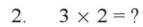

2. $3 \times 2 = ?$

3. Is 933 an even or an odd number?

4. A three-sided shape is called a(n) _____.

5. $525 - 376 = ?$

6. Which digit is in the hundreds place in 36,244?

7. There are 5 shelves of hats in a closet. Each shelf holds 4 hats. How many hats are in the closet?

8. $2,498 + 6,576 = ?$

9. $4 \times 3 = ?$

10. What time was it 15 minutes ago, if it is 6:45 now?

11. If you have seven dimes and four nickels, how much money do you have?

12. Write $5,000 + 300 + 80 + 7$ in standard form.

13. Draw a triangle. Show a line of symmetry.

14. Round 81,574 to the nearest ten thousand.

15. Draw a line.

1.	2.	3.
4.	5.	6.
7.	8.	9.
10.	11.	12.
13.	14.	15.

Lesson #36

1. $8 + 7 = ?$

2.   Draw 2 similar triangles.

3. $800 - 533 = ?$

4.   How many days are in 3 weeks?

5. $4 \times 5 = ?$

6.   Which digit is in the ones place in 3,568?

7.   Samantha has 6 flowerpots. If she plants 3 flowers in each pot, how many flowers will Samantha plant?

8. $5 \times 0 = ?$

9.   42,653 $\bigcirc$ 42,536

10. It is 10:05. What time will it be in 25 minutes?

11. Name the shape.

12. Round 717,256 to the nearest ten thousand.

13. $422 + 657 = ?$

14. Put these numbers in order from greatest to least.

    816   861   618   681

15. How long is the rectangle?

1.	2.	3.
4.	5.	6.
7.	8.	9.
10.	11.	12.
13.	14.	15.

Lesson #37

1. $4 \times 10 = ?$

2. Round 486,255 to the nearest hundred thousand.

3. $379 + 837 = ?$

4. Which digit is in the ten thousands place in 316,275?

5. How many months are in 3 years?

6. If it is 5:25 now, what time was it 15 minutes ago?

7. Draw 2 similar squares.

8. $6,532 - 3,871 = ?$

9. A four-sided shape is a(n) _____.

10. How many dimes are in $3.00?

11. Mrs. Jackson spent 45 minutes at the grocery store. She finished shopping at 1:30 p.m. At what time did she arrive at the store?

12. $3 \times 6 = ?$

13. Is 279 an even or an odd number?

14. $27 + 56 = ?$

15. Write a multiplication sentence for $6 + 6 + 6 + 6$.

1.

2.

3.

4.

5.

6.

7.

8.

9.

10.

11.

12.

13.

14.

15.

Lesson #38

1.   Draw a line.

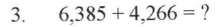

2.   **A line segment is part of a line. It has 2 endpoints.**
     Draw a line segment.

3. $6,385 + 4,266 = ?$

4. $6 \times 2 = ?$

5.   How many feet are in 3 yards?

6. $591 - 388 = ?$

7.   What are the first 5 odd numbers?

8.   Would an adult be about 6 inches tall
     or 6 feet tall?

9. $5 \times 7 = ?$

10. Round 92,561 to the nearest hundred.

11. The answer to a multiplication problem is the _____.

12. Write the next 2 numbers in the sequence. 18, 22, 26, ...

13. $12 + 16 + 27 = ?$

14. 9,753 ◯ 9,537

15. Sherry bought 2 notebooks that cost $1.79 each. She paid the clerk
     with a five-dollar bill. How much change did she receive?

1.	2.	3.
4.	5.	6.
7.	8.	9.
10.	11.	12.
13.	14.	15.

Lesson #39

1. Put these numbers in order from least to greatest.

   3,479   3,736   3,381   3,692

2. $6 \times 5 = ?$

3. The clock shows what time?

4. $700 - 245 = ?$

5. Draw a line segment.

6. The answer to a subtraction problem is the _____.

7. $9,362 + 5,447 = ?$

8. How many minutes are in 2 hours?

9. $3 \times 8 = ?$

10. Karen has 3 quarters, 2 dimes, and 3 pennies. How much money does she have?

11. Round 3,796 to the nearest thousand.

12. How many quarts are in a gallon?

13. Each student had one vote. How many students voted?

14. How many more students voted for ice-cream than for pie?

15. Which is longer, 8 feet or 8 miles?

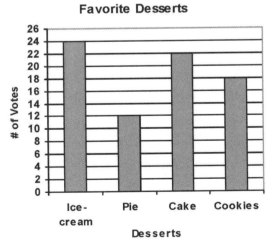

Favorite Desserts

1.	2.	3.
4.	5.	6.
7.	8.	9.
10.	11.	12.
13.	14.	15.

Lesson #40

1. The sum is the answer to a(n) _____ problem.

2. $780 - 592 = ?$

3. Write the even numbers between 21 and 29.

4. $6 \times 7 = ?$

5. How many nickels are in $1.50?

6. $8,937 + 5,891 = ?$

7. How many ounces are in a pound?

8. Draw 2 congruent rectangles.

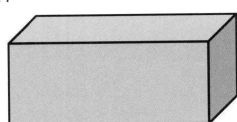

9. $4 \times 8 = ?$

10. What time will it be in 1 hour and 10 minutes, if it is 9:00 now?

11. Write 62,349 in expanded form.

12. Maggie rode her bike 3 miles farther than Gina. Together, they rode 25 miles. How many miles did each girl ride?

13. Is 36,245 an even or an odd number?

14. What is the name of this shape?

15. Write 398 using words.

1.

2.

3.

4.

5.

6.

7.

8.

9.

10.

11.

12.

13.

14.

15.

Lesson #41

1.   Draw a line segment.

2. $6 \times 6 = ?$

3.   Which digit is in the hundreds place in 45,206?

4. $606 - 277 = ?$

5.   Martin has 7 quarters. How much money does he have?

6.   What time is shown on the clock?

7. $3,754 + 9,826 = ?$

8. $3 \times 4 = ?$

9.   How many inches are in a foot?

10.   78,257 $\bigcirc$ 78,527

11.   Name each of the shapes: a) $\triangle$ b) $\square$ c) $\square$

12.   Write the multiplication sentence for $6 + 6 + 6 + 6 + 6 + 6$.

13.   Write the standard number for
      $70,000 + 3,000 + 200 + 8$.

14.   What are the numbers 3 and 5
      called in the multiplication
      sentence $3 \times 5 = 15$?

15.   Joel spent $2.35 for a hamburger and $1.09 for French fries at lunch.
      How much money did Joel spend on lunch?

1.	2.	3.
4.	5.	6.
7.	8.	9.
10.	11.	12.
13.	14.	15.

Lesson #42

1. Jackie is 7 years older than Billy. Billy is 5 years older than Tia. Billy is 7 years old. How old are Jackie and Tia?

2. $2,589 + 8,367 = ?$

3. Draw a line segment.

4. $7 \times 4 = ?$

5. Round 746,322 to the nearest hundred thousand.

6. It is 2:20 now. What time was it 1 hour and 20 minutes ago?

7. $6,125 - 3,776 = ?$

8. If Sean had 6 dimes, 3 nickels, and 4 pennies, how much money did he have?

9. A minute is how many seconds?

10. $73 + 39 + 11 = ?$

11. $7 \times 8 = ?$

12. Would a table be 3 inches tall or 3 feet tall?

13. What number comes before 8,560?

14. Write a multiplication sentence for $7 + 7 + 7$ and solve it.

15. Write *seven thousand, six hundred fifty-six* as a standard number.

1.	2.	3.
4.	5.	6.
7.	8.	9.
10.	11.	12.
13.	14.	15.

Lesson #43

1. $2 \times 9 = ?$

2. A(n) _____ is a four-sided, closed figure.

3. How many quarts are in 2 gallons?

4. Draw a line.

5. $7,353 + 8,649 = ?$

6. Round 92,355 to the nearest ten thousand.

7. $7 \times 5 = ?$

8. Is the number 56,371 even or odd?

9. $7 + 5 = ?$

10. $5,370 - 2,893 = ?$

11. $6 \times 0 = ?$

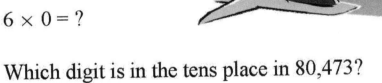

12. Which digit is in the tens place in 80,473?

13. If it is 12:25, what time will it be in 2 hours and 5 minutes?

14. William did his math homework before his social studies. He did his reading homework after his social studies. He did his spelling homework before his math. Which homework did William do last?

15. Is it best to weigh a leaf in ounces or in pounds?

1.	2.	3.
4.	5.	6.
7.	8.	9.
10.	11.	12.
13.	14.	15.

Lesson #44

1. $7 \times 2 = ?$

2.   Round 875,340 to the nearest hundred thousand.

3. $56 + 78 = ?$

4.   Draw a quadrilateral.

5.   How many ounces are in 2 pounds?

6. $4,307 - 2,765 = ?$

7. $5 \times 6 = ?$

8.   What is the answer to a subtraction problem called?

9. $6,399 + 7,805 = ?$

10.   Give the name of the shape.

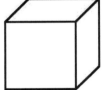

11.   Write $700,000 + 20,000 + 9,000 + 800$ as a standard number.

12.   List the first 4 even numbers.

13.   If it is 7:10 now, what time will it be in 2 hours and 20 minutes?

14. $3 \times 8 = ?$

15.   Stan wants to buy a video game that costs $45.98 and a new CD that costs $16.99. He has $60.00. Does Stan have enough money to buy both the game and the CD?

1.	2.	3.
4.	5.	6.
7.	8.	9.
10.	11.	12.
13.	14.	15.

Lesson #45

1. Draw 2 similar circles.

2. Scott rides his bike 2 miles on Monday, 5 miles on Tuesday, and 8 miles on Wednesday. If this pattern continues, how far will Scott ride his bike on Friday?

3. Write a multiplication sentence for $3 + 3 + 3 + 3 + 3$ and solve it.

4. $448 + 373 = ?$

5. Round 83,526 to the nearest thousand.

6. Draw a square and shade $\dfrac{3}{4}$ of it.

7. $8,127 - 5,632 = ?$

8. $9,325 \bigcirc 9,235$

9. $6 \times 5 = ?$

10. Write *sixteen thousand, seven hundred forty-five* as a standard number.

11. **Any closed figure made up of line segments is called a polygon.** Draw these polygons. a) △ b) ▢ c) ▭

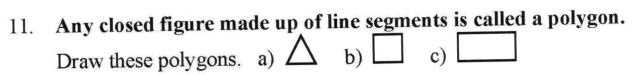

12. $8 \times 3 = ?$

13. What are the odd numbers between 50 and 56?

14. Draw a line segment.

15. How many hours are in a day? In 2 days?

1.	2.	3.
4.	5.	6.
7.	8.	9.
10.	11.	12.
13.	14.	15.

Lesson #46

1.   Draw a 3-sided polygon.

2. $8 \times 7 = ?$

3. $19 + 38 + 52 = ?$

4.   How many quarters are in $3?

5.   The sum is the answer to what kind of problem?

6. $8,314 - 6,884 = ?$

7.   Which digit is in the hundreds place in 37,056?

8.   If it is 9:15 now, what time was it 3 hours and 10 minutes ago?

9. $4 \times 4 = ?$

10. How many books did Marcus read?

11. How many more books did Randy read than Danita?

Number of Books Read	
Tina	📖📖📖
Randy	📖📖📖📖
Danita	📖📖
Marcus	📖📖📖📖📖

Each 📖 stands for 3 books.

12. Round 18,327 to the nearest ten thousand.

13. $8 \times 10 = ?$

14. How many feet are in 3 yards?

15. $397 + 468 = ?$

1.	2.	3.
4.	5.	6.
7.	8.	9.
10.	11.	12.
13.	14.	15.

Lesson #47

1. Julia has 3 boxes of pencils with 4 pencils in each box. How many pencils does Julia have?

2. Is 17,314 an even or an odd number?

3. 864,376 $\bigcirc$ 846,763

4. Look at this clock. What time is it?

5. 18 − 9 = ?

6. 6,302 − 3,947 = ?

7. How many inches are in a foot?

8. 5 × 9 = ?

9. A closed shape made up of line segments is called a(n) _____.

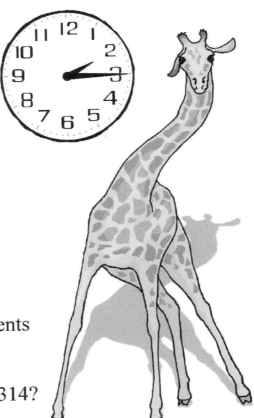

10. Which digit is in the tens place in 27,314?

11. 6 × ? = 42

12. 998 + 764 = ?

13. Write the next number in the sequence. 23, 30, 37, ...

14. Round 876 to the nearest hundred.

15. Tony has a five-dollar bill, 3 one-dollar bills, 2 quarters, and 2 dimes. How much money does he have?

1.	2.	3.
4.	5.	6.
7.	8.	9.
10.	11.	12.
13.	14.	15.

Lesson #48

1. I am a polygon with 4 equal sides. What am I?

2. $8 \times 9 = $?

3. How many quarts are in a gallon?

4. Draw a line.

5. $7,739 - 2,863 = $?

6. _____ multiplied by any number is equal to zero.

7. Would a bracelet weigh about 12 ounces or 12 pounds?

8. $3 \times 9 = $?

9. What fraction is shaded?

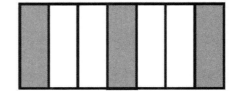

10. Mr. Jasper needs to drive his delivery to its destination 2,195 miles away. He has already driven 988 miles. How many more miles does Mr. Jasper have to drive to reach his destination?

11. $5,931 + 7,687 = $?

12. Draw 2 congruent rectangles.

13. $32 + 48 + 11 = $?

14. $5 \times 8 = $?

15. It is 1:05 right now. What time will it be in 4 hours and 10 minutes?

1.	2.	3.
4.	5.	6.
7.	8.	9.
10.	11.	12.
13.	14.	15.

Lesson #49

1. $9 \times 6 = ?$

2. Which digit is in the thousands place in 48,203?

3. How many seconds are in a minute?

4. $5,953 + 8,319 = ?$

5. Round 132,566 to the nearest hundred thousand.

6. $602 - 355 = ?$

7. $3 \times 9 = ?$

8. Draw a line segment.

9. The answer to a multiplication problem is called the _____.

10. I am a polygon with 3 sides. What am I?

11. $2 \times 8 = ?$

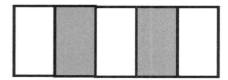

12. What fraction is shaded?

13. What time was it 3 hours and 5 minutes ago, if it is 3:20 now?

14. Todd had 3 baskets with 8 apples in each basket. How many apples did Todd have?

15. Write 63,214 in expanded form.

1.

2.

3.

4.

5.

6.

7.

8.

9.

10.

11.

12.

13.

14.

15.

Lesson #50

1. Draw 2 similar hearts.

2. $8,314 - 5,733 = ?$

3. $9 \times 9 = ?$

4. Draw a square and a line of symmetry through the square.

5. $57,331 \bigcirc 53,713$

6. Write a multiplication sentence for $7 + 7 + 7 + 7$ and solve it.

7. There are 5 spiders on the wall. Each spider has 8 legs. How many spider legs are on the wall?

8. How many sides does a quadrilateral have?

9. $6,313 + 9,254 = ?$

10. $8 \times 7 = ?$

11. Write the time.

12. How much money does Shirley have if she has 5 quarters and 3 dimes?

13. Would a fence be about 5 feet tall or 5 inches tall?

14. Write the even numbers between 41 and 49.

15. $26 + 33 + 12 = ?$

1.

2.

3.

4.

5.

6.

7.

8.

9.

10.

11.

12.

13.

14.

15.

Lesson #51

1. Round 237,418 to the nearest hundred.

2. $5 \times 5 = ?$

3. **A five-sided polygon is called a pentagon.**
   Draw a pentagon.

4. $9,326 - 3,897 = ?$

5. How many days are in a year?

6. $2 \times 8 = ?$

7. It is 1:00 now. What time will it be
   in 4 hours and 20 minutes?

8. $9 + 4 = ?$

9. What is the name for a part of a line?

10. Which digit is in the thousands place in 4,862?

11. Write the next number in the sequence. 15, 22, 29, ...

12. Mrs. Yates made 5 costumes for the school play. On each costume
    she sewed 4 ribbons. How many ribbons did she sew in all?

13. Put these numbers in order from greatest to least.
    4,076   4,983   4,112   4,362

14. $462 + 597 = ?$

15. In a multiplication problem, the two numbers that are multiplied
    together are the _____.

1.

2.

3.

4.

5.

6.

7.

8.

9.

10.

11.

12.

13.

14.

15.

Lesson #52

1.   Draw a pentagon.

2. $6 \times 6 = ?$

3. $807 - 355 = ?$

4.   Write $70{,}000 + 5{,}000 + 300 + 8$ as a standard number.

5.   In a dog walk, Fluffy is behind Kiely. Muffin is ahead of Thor. Muffin is between Fluffy and Thor. What is the order of the dogs?

6.   Round 136,812 to the nearest ten thousand.

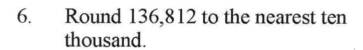

7. $3{,}874 + 9{,}216 = ?$

8.   Write 8,136 using words.

9. $96 - 37 = ?$

10.   What time will it be in 3 hours, if it is 5:30 now?

11. $4 \times 3 = ?$

12.   How many inches are in a foot?

13. $14 - 6 = ?$

14.   Figures with the same size and shape are _____ .

15.   Write the odd numbers between 82 and 88.

1.	2.	3.
4.	5.	6.
7.	8.	9.
10.	11.	12.
13.	14.	15.

Lesson #53

1. Which is longer, 6 inches or 1 foot?

2. $715 - 266 = ?$

3. The difference is the answer to what type of problem?

4. Draw a line.

5. How many dimes are in 4 dollars?

6. What time is shown on the clock?

7. Draw a rectangle. Shade $\frac{2}{6}$ of it.

8. $3 \times 3 = ?$

9. $3,994 + 6,337 = ?$

10. Write the next number in the sequence. 30, 35, 40, ...

11. I have four sides. Each of my sides is the same length. Draw me.

12. What is the name of this shape?

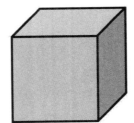

13. Round 16,344 to the nearest ten.

14. $5 \times 8 = ?$

15. What is the length of the stick?

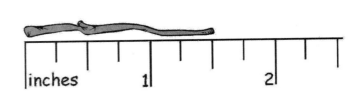

1.	2.	3.
4.	5.	6.
7.	8.	9.
10.	11.	12.
13.	14.	15.

Lesson #54

1. I am a polygon with 5 sides. What am I?

2. $5,816 - 2,487 = ?$

3. $7 \times 5 = ?$

4. Which digit is in the hundreds place in 74,062?

5. How many quarts are in a gallon?

6. Write an addition sentence for 5×4.

7. $8,458 + 4,298 = ?$

8. Round 394,556 to the nearest thousand.

9. $8 \times 8 = ?$

10. What time was it 2 hours and 15 minutes ago, if it is 6:45 now?

11. Any number multiplied by zero is equal to _____.

12. $13 - 8 = ?$

13. $2 \times 7 = ?$

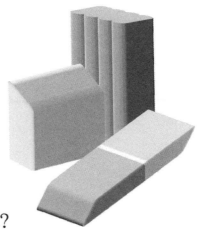

14. Would a dictionary best be weighed in ounces or in pounds?

15. Sam bought 6 pencils that cost $0.07 each and 4 erasers that cost $0.04 each. How much money did Sam spend on his supplies?

1.	2.	3.
4.	5.	6.
7.	8.	9.
10.	11.	12.
13.	14.	15.

Lesson #55

1. Write 27,356 in expanded form.

2. $16 + 28 + 10 = ?$

3. How many months are in two years?

4. $7,325 - 5,864 = ?$

5. $9 \times 6 = ?$

6. Figures having the same shape, but different sizes are _____.

7. $347 + 996 = ?$

8. $3 \times 8 = ?$

9. $9 + 8 = ?$

10. Draw a pentagon.

11. 8,746 ◯ 8,476

12. How many nickels are in a dollar?

13. What are the even numbers between 31 and 39?

14. Ramon went to the park at 12:15 p.m. He left the park 2 hours and 15 minutes later. At what time did Ramon leave the park?

15. If it is 7:10 now, what time will it be in 2 hours and 10 minutes?

1.

2.

3.

4.

5.

6.

7.

8.

9.

10.

11.

12.

13.

14.

15.

Lesson #56

1.   Write the next 2 numbers in the sequence.   52,   60,   68, ...

2. $96 - 48 =$?

3. $9 \times 9 =$?

4.   Write *twenty-six thousand, nine hundred thirty-six* as a standard number.

5. $2,887 + 7,925 =$?

6.   Which digit is in the ten thousands place in 307,815?

7. $4 \times 5 =$?

8.   Each year in a human's life is said to equal 7 years in a dog's life. If a dog is 4 human-years old, what is its age in dog-years?

9.   How many feet are in a yard?

10. $8 \times 10 =$?

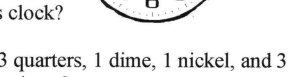

11. What time is it, according to this clock?

12. Nicholas had 4 one-dollar bills, 3 quarters, 1 dime, 1 nickel, and 3 pennies. How much money did he have?

13. Draw a pentagon.

14. Draw a heart. Show a line of symmetry on the heart.

15. Round 486 to the nearest hundred.

1.	2.	3.
4.	5.	6.
7.	8.	9.
10.	11.	12.
13.	14.	15.

Lesson #57

1. The recycling center pays 4¢ for each can. Cory brought 9 cans to the recycling center. How much money will he receive?

2. Draw a pentagon.

3. Which digit is in the hundred thousands place in 862,354?

4. $507 - 269 = ?$

5. $56 \times 2 = ?$

6. It is 7:45 now. What time will it be in 5 hours and 15 minutes?

7. Write 36,240 in expanded form.

8. $8 \times 9 = ?$

9. **There are 2 cups in a pint.** Write *2 cups = 1 pint* three times.

10. $7,431 + 8,669 = ?$

11. Round 372,564 to the nearest hundred.

12. What fraction is shaded?

13. What is the answer to a multiplication problem called?

14. $3 \times 8 = ?$

15. How many inches are in 3 feet?

1.

2.

3.

4.

5.

6.

7.

8.

9.

10.

11.

12.

13.

14.

15.

Lesson #58

1. $319 + 665 = ?$

2. Which is the greater length, 2 feet or 1 yard?

3. How many quarters are in $4?

4. Draw 2 congruent pentagons.

5. Adam saw 7 cars with 4 passengers in each. How many people did Adam see in all?

6. If it is 8:55 now, what time was it 3 hours and 10 minutes ago?

7. $8,000 - 3,543 = ?$

8. A four-sided polygon is called a(n) _____.

9. $32 \times 3 = ?$

10. Draw a line segment.

11. $56,298 \bigcirc 56,982$

12. How many cups are in a pint?

13. $8,367 + 8,447 = ?$

14. $5 \times 9 = ?$

15. Write the odd numbers between 80 and 88.

1.

2.

3.

4.

5.

6.

7.

8.

9.

10.

11.

12.

13.

14.

15.

Lesson #59

1. What is a closed figure that is made up of line segments called?

2. $45 \times 5 = ?$

3. Zero multiplied by any number is equal to what?

4. A five-sided polygon is called a(n) _____.

5. $517 - 398 = ?$

6. Nicole spends $6 per week to ride her horse. How much does Nicole spend to ride her horse for 6 weeks?

7. $6 \times 4 = ?$

8. Draw a square. Show 2 lines of symmetry.

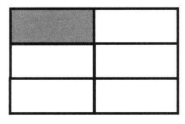

9. $2 \times 10 = ?$

10. How many ounces are in 2 pounds?

11. Write $5,000 + 400 + 90 + 6$ as a standard number.

12. Round 475,619 to the nearest thousand.

13. Put these numbers in order from least to greatest.

    4,365   4,635   4,356   4,653

14. What fraction of this figure is <u>not</u> shaded?

15. Would the length of a highway best be measured in feet or in miles?

1.

2.

3.

4.

5.

6.

7.

8.

9.

10.

11.

12.

13.

14.

15.

Lesson #60

1. Draw a line.

2. $3 \times 8 = ?$

3. $19 + 28 + 36 = ?$

4. It is 7:25 now. What time will it be in 2 hours and 5 minutes?

5. Round 16,257 to the nearest thousand.

6. Jacob bought a hot dog for $1.75 and a soft drink for $1.09. He gave the clerk $5. How much change did he receive?

7. How many quarts are in four gallons?

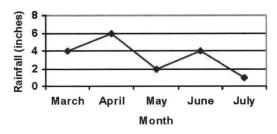

8. $8,305 - 4,416 = ?$

9. Which two months had the same amount of rainfall?

10. During which month did the least amount of rain fall?

11. The answer to a subtraction problem is called the _____.

12. $5 \times 10 = ?$

13. Would a strawberry weigh about 1 ounce or 1 pound?

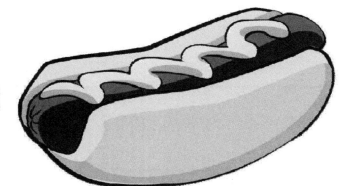

14. $7 \times 9 = ?$

15. $3,559 + 788 = ?$

1.

2.

3.

4.

5.

6.

7.

8.

9.

10.

11.

12.

13.

14.

15.

Lesson #61

1. The answer to an addition problem is called the _____.

2. $6 \times 9 = ?$

3. Would a bowling ball best be weighed in ounces or in pounds?

4. $6,000 - 2,431 = ?$

5. How many cups are in a pint?

6. Which digit is in the tens place in 47,302?

7. $6,345 \bigcirc 6,453$

8. $36 \div 4 = ?$

9. What is the name of this shape?

10. How many hours are in 3 days?

11. $37 \times 4 = ?$

12. What will be the time 20 minutes before 3:00?

13. Patty has 2 coins that total $0.35. What are the coins?

14. Mrs. Harris has 5 grocery bags full of bread. In each grocery bag are 5 loaves of bread. How much bread does Mrs. Harris have?

15. Write the missing numbers in the sequence. 36, 40, ____, 48, 52, ____

1.

2.

3.

4.

5.

6.

7.

8.

9.

10.

11.

12.

13.

14.

15.

Lesson #62

1. 6,314 + 8,552 = ?

2. 3 × 7 = ?

3. Round 28,076 to the nearest ten thousand.

4. Write 3,475 using words.

5. 25 ÷ 5 = ?

6. Is 371,215 an even or an odd number?

7. 923 − 677 = ?

8. How many cups are in a pint?

9. Pete gathered 4,568 pounds of walnuts and 3,216 pounds of pecans. How many pounds of nuts did Pete gather in all?

10. Draw 2 similar squares.

11. 6 × 7 = ?

12. If you have fifty nickels, how much money do you have?

13. (2 × 3) × 2 = ?

14. 63 × 5 = ?

15. Jamal's baseball glove weighs 14 ounces. How much less than 1 pound does his glove weigh?

1.	2.	3.
4.	5.	6.
7.	8.	9.
10.	11.	12.
13.	14.	15.

Lesson #63

1. $2 \times 9 = ?$

2. A quotient is an answer to what type of problem?

3. What fraction of the circle is shaded?

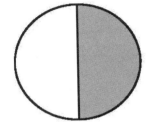

4. $40 \div 8 = ?$

5. How many days are in 7 weeks?

6. $3,699 + 9,988 = ?$

7. Would it be more likely for a horse to weigh 700 ounces or 700 pounds?

8. $6 \times 6 = ?$

9. Which is greater, $(50,000 + 2,000 + 700 + 90 + 2)$ or $53,987$?

10. $32 \div 4 = ?$

11. Which digit is in the hundreds place in 76,405?

12. $712 - 486 = ?$

13. $25 \times 6 = ?$

14. What are the even numbers between 41 and 45?

15. What is the date one week before April 28th? On which day of the week is May 1st?

1.	2.	3.
4.	5.	6.
7.	8.	9.
10.	11.	12.
13.	14.	15.

Lesson #64

1. Jenna baked 2 dozen cupcakes. She divided the cupcakes between herself and 3 of her friends. How many cupcakes did each one get?

2. $54 \div 9 = ?$

3. $600 - 253 = ?$

4. **A ray is part of a line. It has only one end point.** Draw a ray.

5. $5 \times 3 = ?$

6. The product is an answer to a(n) _____ problem.

7. Round 478,351 to the nearest hundred thousand.

8. Look at this clock. What time is it?

9. $30 \div 6 = ?$

10. William has 2 five-dollar bills, 3 one-dollar bills, 1 quarter, 4 dimes, and a nickel. How much money does he have?

11. Draw a line segment.

12. $6 \times 9 = ?$

13. Write 3,876 in expanded form.

14. $4,308 + 9,667 = ?$

15. I am a solid shape with 6 faces. All of my faces are squares. What am I?

1.	2.	3.
4.	5.	6.
7.	8.	9.
10.	11.	12.
13.	14.	15.

Lesson #65

1. Ann received four coins in change that added up to 37¢. What coins did she receive?

2. $7 \times 8 = ?$

3. $7,435 - 2,982 = ?$

4. Write the next number in the sequence.
   55, 63, 71, …

5. How many cookies are in 4 dozen?

6. $362 + 418 + 173 = ?$

7. $16 \div 4 = ?$

8. Round 773,219 to the nearest ten.

9. $2 \times 6 = ?$

10. What time was it 3 hours and 15 minutes ago, if it is 11:45 now?

11. $56 \times 3 = ?$

12. A quadrilateral has _____ sides.

13. Draw a ray.

14. A baseball bat weighs 1 pound and 9 ounces. What is its weight in ounces?

15. The answer to an addition problem is called the _____.

1.	2.	3.
4.	5.	6.
7.	8.	9.
10.	11.	12.
13.	14.	15.

Lesson #66

1. **There are 2 pints in a quart.** Write *2 pints = 1 quart* three times.

2. $49 \div 7 = ?$

3. $8,312 - 5,705 = ?$

4. How many feet are in four yards?

5. A closed figure made up of line segments is called a(n) _____.

6. $4 \times 8 = ?$

7. Write the time.

8. Marissa saw 8 bees in the garden. Each bee had 2 wings. How many wings did she count?

9. $807 + 983 = ?$

10. $25 \times 7 = ?$

11. Any number multiplied by one is equal to _____.

12. What fraction of the figure is shaded?

13. How many cups are in a pint?

14. Round $8.21 to the nearest dollar.

15. Draw a ray.

1.

2.

3.

4.

5.

6.

7.

8.

9.

10.

11.

12.

13.

14.

15.

Lesson #67

1. $9 \times 8 = ?$

2. How many inches are in 5 feet?

3. Round 31,214 to the nearest hundred.

4. $27 \div 3 = ?$

5. An answer to a division problem is called the _____.

6. $4,256 + 7,773 = ?$

7. $3 \times 6 = ?$

8. Draw a line.

9. $5,122 - 2,686 = ?$

10. $27 \times 4 = ?$

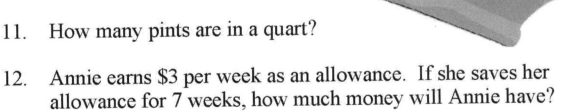

11. How many pints are in a quart?

12. Annie earns $3 per week as an allowance. If she saves her allowance for 7 weeks, how much money will Annie have?

13. Draw 2 congruent pentagons.

14. Does a bathtub filled with water hold about 15 pints or 15 gallons?

15. If it is 3:30 now, what time was it 1 hour and 30 minutes ago?

1.

2.

3.

4.

5.

6.

7.

8.

9.

10.

11.

12.

13.

14.

15.

Lesson #68

1. $32 + 45 + 13 = ?$

2. Which digit is in the ten thousands place in 531,076?

3. $7 \times 5 = ?$

4. What will be the time 20 minutes before 11:00?

5. Would an aquarium hold 10 cups or 10 gallons of water?

6. $81 \div 9 = ?$

7. How many sides does a pentagon have?

8. $53 \times 6 = ?$

9. Draw a ray.

10. $9,000 - 2,774 = ?$

11. Write $600,000 + 90,000 + 800 + 70 + 1$ as a standard number.

12. How many ounces are in 3 pounds?

13. $60 \div 6 = ?$

14. $6,347 + 8,999 = ?$

15. Karen earned $15 in one week of baby-sitting. The next week she earned $8. How much did she earn altogether?

1.	2.	3.
4.	5.	6.
7.	8.	9.
10.	11.	12.
13.	14.	15.

Lesson #69

1. How many pints are in a quart?

2. Chad was 20 minutes late for soccer practice. If practice began at 6:15 p.m., at what time did Chad arrive for practice?

3. $28 \div 4 = ?$

4. Write 142,560 in expanded form.

5. It is 6:40 now. What time will it be in 5 hours and 20 minutes?

6. What fraction is shaded?

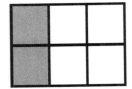

7. $2,797 + 9,884 = ?$

8. Name the following shapes. a) ☐ b) ⬠ c) △

9. Round 4,143 to the nearest hundred.

10. $73 \times 8 = ?$

11. Draw a line segment.

12. Write the next number in the sequence. 47, 50, 53, ...

13. $5 \times 8 = ?$

14. A four-sided polygon is called a _____.

15. $76 - 29 = ?$

1.	2.	3.
4.	5.	6.
7.	8.	9.
10.	11.	12.
13.	14.	15.

Lesson #70

1. $44 \times 4 = ?$

2. $7 \times 6 = ?$

3. Round 148 to the nearest hundred.

4. $12 \div 4 = ?$

5. $63 \div 7 = ?$

6. How many feet are in 5 yards?

7. The answer to a subtraction problem is called the _____.

8. Any number multiplied by zero is equal to _____.

9. $5,567 + 9,832 = ?$

10. Draw a ray.

11. Tickets to the play cost $15 for adults and $7 for children. How much will it cost for a family of 2 adults and 2 children to go to the play?

12. According to the clock, what is the time?

13. How many cups are in a pint?

14. $9,216 - 6,377 = ?$

15. $237 \times 3 = ?$

1.	2.	3.
4.	5.	6.
7.	8.	9.
10.	11.	12.
13.	14.	15.

Lesson #71

1. $67 \times 5 = ?$

2. How many months pass from July 1ˢᵗ to December 1ˢᵗ?

3. $55 + 88 = ?$

4. What will be the time 50 minutes after 6:00?

5. How many months are in 3 years?

6. The animal trainer is feeding some porpoises 3 buckets of fish. Each bucket contains 8 fish. How many fish do the porpoises eat?

7. Round 66,739 to the nearest hundred.

8. $53 \div 7 = ?$ (Hint: There will be a remainder.)

9. Rondell has 7 quarters, 3 dimes, and 3 nickels. How much money does he have?

10. Draw 2 similar circles.

11. $2,075 - 988 = ?$

12. $64 \div 8 = ?$

13. $86,413 \bigcirc 86,143$

14. List the odd numbers between 32 and 38.

15. $274 \times 6 = ?$

1.	2.	3.
4.	5.	6.
7.	8.	9.
10.	11.	12.
13.	14.	15.

Lesson #72

1. $8,817 + 4,566 = ?$

2. How many pints are in a quart?

3. $139 \times 7 = ?$

4. What will be the time 10 minutes before midnight?

5. $49 \div 6 = ?$

6. Draw a ray.

7. $822 - 657 = ?$

8. A pentagon has _____ sides.

9. $8 \times 7 = ?$

10. Put these numbers in order from greatest to least.

    9,325   9,532   9,247   9,427

11. $70 \div 10 = ?$

12. Is 72,480 an even or an odd number?

13. There were 561 people at the concert and 387 people at the opera. How many more people attended the concert than the opera?

14. Name the shape.

15. What do we call figures with the same size and shape?

1.	2.	3.
4.	5.	6.
7.	8.	9.
10.	11.	12.
13.	14.	15.

Level 3

Mathematics

Help Pages

Help Pages

Vocabulary

Arithmetic Operations
Difference — the result or answer to a subtraction problem. Example: The difference of 5 and 1 is 4.
Product — the result or answer to a multiplication problem. Example: The product of 5 and 3 is 15.
Quotient — the result or answer to a division problem. Example: The quotient of 8 and 2 is 4.
Sum — the result or answer to an addition problem. Example: The sum of 5 and 2 is 7.

Geometry
Acute Angle — an angle measuring less than 90°.
Area — the size of a surface. Area is always given in square units (feet^2, meters^2,...).
Congruent — figures with the same shape and the same size.
Denominator — the bottom number of a fraction. Example: $\frac{1}{4}$ ➡ denominator is 4
Diameter — the widest distance across a circle. The diameter always passes through the center.
Fraction — a part of a whole. Example: ⊞ This box has 4 parts. 1 part is shaded. $\frac{1}{4}$
Line of Symmetry — a line along which a figure can be folded so that the two halves match exactly.
Numerator — the top number of a fraction. Example: $\frac{1}{4}$ ➡ numerator is 1
Obtuse Angle — an angle measuring more than 90°.
Perimeter — the distance around the outside of a polygon.
Radius — the distance from any point on the circle to the center. The radius is half of the diameter.
Remainder — the part left over when one number can't be divided exactly by another.
Right Angle — an angle measuring exactly 90°.
Similar — figures having the same shape, but different sizes.

Geometry — Polygons

Number of Sides		Name	Number of Sides		Name
3	△	Triangle	6	⬡	Hexagon
4	☐	Quadrilateral	8	⯃	Octagon
5	⬠	Pentagon			

Help Pages

Vocabulary

Measurement — Relationships	
Volume	**Distance**
3 teaspoons in a tablespoon	36 inches in a yard
2 cups in a pint	1760 yards in a mile
2 pints in a quart	5280 feet in a mile
4 quarts in a gallon	100 centimeters in a meter
Weight	1000 millimeters in a meter
16 ounces in a pound	**Temperature**
2000 pounds in a ton	0° Celsius – Freezing Point
Time	100° Celsius – Boiling Point
10 years in a decade	32° Fahrenheit – Freezing Point
100 years in a century	212° Fahrenheit – Boiling Point

Statistics

Mode — the number that occurs most often in a group of numbers. The mode is found by counting how many times each number occurs in the list. The number that occurs more than any other is the mode. Some groups of numbers have more than one mode.

Example: The mode of 77, ⑨③, 85, ⑨③, 77, 81, ⑨③ and 71 is **93**.
(93 is the mode because it occurs more than the others.)

Place Value

Whole Numbers					
2	7	1,	4	0	5
Hundred Thousands	Ten Thousands	Thousands	Hundreds	Tens	Ones

The number above is read: two hundred seventy-one thousand, four hundred five.

Help Pages

Solved Examples

Whole Numbers (continued)

When we **round numbers**, we are estimating them. This means we focus on a particular place value, and decide if that digit is closer to the next highest number (round up) or to the next lower number (keep the same). It might be helpful to look at the place-value chart on page 285.

Example: Round 347 to the tens place.

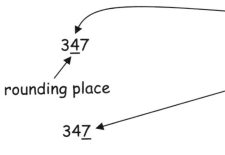

347

rounding place

347

Since 7 is greater than 5, the rounding place is <u>increased by 1</u>.

350

1. Identify the place that you want to round to. What number is in that place? (4)

2. Look at the digit to its right. (7)

3. If this digit is 5 or greater, increase the number in the rounding place by 1. (round up) If the digit is less than 5, keep the number in the rounding place the same.

4. Replace all digits to the right of the rounding place with zeroes.

Here is another example of rounding whole numbers.

Example: Round 4,826 to the hundreds place.

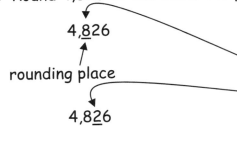

4,826

rounding place

4,826

Since 2 is less than 5, the rounding place <u>stays the same</u>.

4,800

1. Identify the place that you want to round to. What number is in that place? (8)

2. Look at the digit to its right.

3. If this digit is 5 or greater, increase the number in the rounding place by 1. (round up) If the digit is less than 5, keep the number in the rounding place the same.

4. Replace all digits to the right of the rounding place with zeroes.

Help Pages

Solved Examples

Whole Numbers (continued)

When adding or subtracting whole numbers, first the numbers must be lined-up on the right. Starting with the ones place, add (or subtract) the numbers; when adding, if the answer has 2 digits, write the ones digit and regroup the tens digit (for subtraction, it may also be necessary to regroup first). Then, add (or subtract) the numbers in the tens place. Continue with the hundreds, etc.

Look at these examples of **addition**.

Examples: Find the sum of 314 and 12.

Add 6,478 and 1,843.

$$\begin{array}{r} 314 \\ +12 \\ \hline 326 \end{array}$$

1. Line up the numbers on the right.
2. Beginning with the ones place, add. Regroup if necessary.
3. Repeat with the tens place.
4. Continue this process with the hundreds place, etc.

$$\begin{array}{r} {}^{1}\ {}^{1}\ {}^{1}\ \\ 6,478 \\ +1,843 \\ \hline 8,321 \end{array}$$

Use the following examples of **subtraction** to help you.

Examples: Subtract 37 from 93.

$$\begin{array}{r} {}^{8}\ {}^{13} \\ \cancel{9}\ \cancel{3} \\ -37 \\ \hline 56 \end{array}$$

1. Begin with the ones place. Check to see if you need to regroup. Since 7 is larger than 3, you must regroup to 8 tens and 13 ones.
2. Now look at the tens place. Since 3 is less than 8, you do not need to regroup.
3. Subtract each place value beginning with the ones.

Find the difference of 425 and 233.

$$\begin{array}{r} {}^{3}\ {}^{12} \\ \cancel{4}\ \cancel{2}\ 5 \\ -233 \\ \hline 192 \end{array}$$

1. Begin with the ones place. Check to see if you need to regroup. Since 3 is less than 5, you do not need to regroup.
2. Now look at the tens place. Check to see if you need to regroup. Since 3 is larger than 2, you must regroup to 3 hundreds and 12 tens.
3. Now look at the hundreds place. Since 2 is less than 3, you are ready to subtract.
4. Subtract each place value beginning with the ones.

Help Pages

Solved Examples

Whole Numbers (continued)

Sometimes when doing subtraction, you must **subtract from zero**. This always requires regrouping. Use the examples below to help you.

Examples: Subtract 261 from 500.

$$\begin{array}{r} {\scriptstyle 4\ \ \overset{9}{\cancel{10}}\ \ 10} \\ \cancel{5}\,\cancel{0}\,\cancel{0} \\ -\ 2\ 6\ 1 \\ \hline 2\ 3\ 9 \end{array}$$

1. Begin with the ones place. Since 1 is less than 0, you must regroup. You must continue to the hundreds place, and then begin regrouping.
2. Regroup the hundreds place to 4 hundreds and 10 tens.
3. Then, regroup the tens place to 9 tens and 10 ones.
4. Finally, subtract each place value beginning with the ones.

Find the difference between 600 and 238.

$$\begin{array}{r} {\scriptstyle 5\ \ \overset{9}{\cancel{10}}\ \ 10} \\ \cancel{6}\,\cancel{0}\,\cancel{0} \\ -\ 2\ 3\ 8 \\ \hline 3\ 6\ 2 \end{array}$$

Multiplication is a quicker way to add groups of numbers. The sign (×) for multiplication is read "times." The answer to a multiplication problem is called the product. Use the examples below to help you understand multiplication.

Examples: 3 × 5 is read "three times five."

It means *3 groups of 5* or 5 + 5 + 5.

3 × 5 = 5 + 5 + 5 = 15

The product of 3 × 5 is **15**.

4 × 7 is read "four times seven."

It means *4 groups of 7* or 7 + 7 + 7 + 7.

4 × 7 = 7 + 7 + 7 + 7 = 28

The product of 4 × 7 is **28**.

Help Pages

Solved Examples

Whole Numbers (continued)

It is very important that you memorize your **multiplication facts**. This table will help you, but only until you memorize them!

To use this table, choose a number in the top gray box and multiply it by a number in the left gray box. Follow both with your finger (down and across) until they meet. The number in that box is the product.

An example is shown for you: $2 \times 3 = 6$

×	0	1	2	3	4	5	6	7	8	9	10
0	0	0	0	0	0	0	0	0	0	0	0
1	0	1	2	3	4	5	6	7	8	9	10
2	0	2	4	6	8	10	12	14	16	18	20
3	0	3	6	9	12	15	18	21	24	27	30
4	0	4	8	12	16	20	24	28	32	36	40
5	0	5	10	15	20	25	30	35	40	45	50
6	0	6	12	18	24	30	36	42	48	54	60
7	0	7	14	21	28	35	42	49	56	63	70
8	0	8	16	24	32	40	48	56	64	72	80
9	0	9	18	27	36	45	54	63	72	81	90
10	0	10	20	30	40	50	60	70	80	90	100

Help Pages

Solved Examples

Whole Numbers (continued)

When **multiplying multi-digit whole numbers**, it is important to know your multiplication facts. Follow the steps and the examples below.

Examples: Multiply 23 by 5.

$$\overset{1}{2}3$$
$$\times 5$$
$$\overline{115}$$

$3 \times 5 = 15$ ones or 1 ten and 5 ones

$2 \times 5 = 10$ tens + 1 ten (regrouped) or 11 tens.

1. Line up the numbers on the right.
2. Multiply the digits in the ones place. Regroup if necessary.
3. Multiply the digits in the tens place. Add any regrouped tens.
4. Repeat step 3 for the hundreds place, etc.

Find the product of 314 and 3.

$$3\overset{1}{1}4$$
$$\times \ \ 3$$
$$\overline{942}$$

$4 \times 3 = 12$ ones or 1 ten and 2 ones.

$1 \times 3 = 3$ tens + 1 ten (regrouped) or 4 tens.

$3 \times 3 = 9$ hundreds.

Division is the opposite of multiplication. The symbols for division are ÷ and $\overline{)}$ and are read "divided by." The answer to a division problem is called the quotient.

Remember that multiplication is a way of adding groups to get their total. Think of division as the reverse of this. In a division problem you already know the total and the number in each group. You want to know how many groups there are. Follow the examples below.

Examples: Find the quotient of 12 ÷ 3.   (12 items divided into groups of 3)

The total number is 12.

Each group contains 3.

How many groups are there?   There are 4 groups.

$$12 \div 3 = \mathbf{4}$$

Divide 10 by 2.   (10 items divided into groups of 2)

The total number is 10.

Each group contains 2.

How many groups are there?   There are 5 groups.

$$10 \div 2 = \mathbf{5}$$

Help Pages

Solved Examples

Whole Numbers (continued)

Sometimes when you are dividing, there are items left over that do not make a whole group. These left-over items are called the **remainder**. When this happens, we say that "the whole <u>cannot be divided evenly</u> by that number."

Example: What is 16 divided by 5? (16 items divided into groups of 5)

The total number is 16.

Each group contains 5.

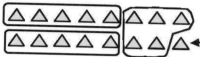

How many groups are there? There are 3 groups, but there is <u>1 left over</u>. The <u>remainder</u> is 1.

$16 \div 5 = $ **3 R1** (This is read "3 remainder 1.")

The next group of examples involves **long division using one-digit divisors with remainders.** You already know how to divide single-digit numbers. This process helps you to be able to divide numbers with multiple digits.

Example: Divide 37 by 4.

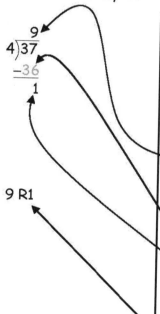

1. In this problem, 37 is the dividend and 4 is the divisor. You're going to look at each digit in the dividend, starting on the left.

2. Ask yourself if the divisor (4) goes into the left-most digit in the dividend (3). It doesn't, so keep going to the right.

3. Does the divisor (4) go into the two left-most digits (37)? It does. How many times does 4 go into 37? (9 times)

4. Multiply 4×9 (product = 36).

5. Subtract 36 from 37 (difference = 1). There's nothing left to bring down from above. Once this number is smaller than the divisor, it is called the remainder and the problem is finished. The remainder is 1.

6. Write the answer (above the top line) with the remainder. (9 R1)

Help Pages

Solved Examples

Whole Numbers (continued)

Example: What is 556 divided by 6?

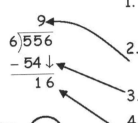

1. Ask yourself if the divisor (6) goes into the left-most digit in the dividend (5). It doesn't, so keep going to the right.

2. Does the divisor (6) go into the two left-most digits (55)? It does. How many times does 6 go into 55? (9 times)

3. Multiply 6 × 9 (product is 54).

4. Subtract 54 from 55. (1) Bring down the 6 ones from the first line. This leaves 16 left from the original 556.

5. Ask yourself if the divisor (6) goes into 16. It does. How many times does 6 go into 16? (2)

6. Multiply 6 × 2 (product is 12).

7. Subtract 12 from 16 (result is 4). There's nothing left to bring down from above. Once this number is smaller than the divisor, it is called the remainder and the problem is finished. The remainder is 4.

8. Write the answer with the remainder. (92 R 4)

Remember: The remainder can NEVER be larger than the divisor!

Fractions

A **fraction** is used to represent part of a whole. The top number in a fraction is called the **numerator** and represents the part. The bottom number in a fraction is called the **denominator** and represents the whole.

The whole rectangle has 6 sections.

Only 1 section is shaded.

This can be shown as the fraction $\frac{1}{6}$.

$$\frac{1}{6} \quad \begin{array}{l} \text{shaded part (numerator)} \\ \text{total parts (denominator)} \end{array}$$

To **add (or subtract) fractions with the same denominator**, simply add (or subtract) the numerators, keeping the same denominator.

Examples: $\frac{3}{5} + \frac{1}{5} = \frac{4}{5}$ $\frac{8}{9} - \frac{1}{9} = \frac{7}{9}$

Help Pages

Solved Examples

Decimals

Adding and subtracting decimals is very similar to adding or subtracting whole numbers. The main difference is that you have to line-up the decimal points in the numbers before you begin. Add zeros if necessary, so that all of the numbers have the same number of digits after the decimal point. Before you subtract, remember to check to see if you must regroup. When you're finished adding (or subtracting), bring the decimal straight down into your answer.

Example: Find the sum of 4.25 and 2.31.

$$
\begin{array}{r}
4.25 \\
+\ 2.31 \\
\hline
6.56
\end{array}
$$

> 1. Line up the decimal points. Add zeroes as needed.
> 2. Add (or subtract) the decimals.
> 3. Add (or subtract) the whole numbers.
> 4. Bring the decimal point straight down.

Example: Subtract 4.8 from 7.4.

$$
\begin{array}{r}
\overset{6\ \ \ 14}{7.\cancel{4}} \\
-\ 4.8 \\
\hline
2.6
\end{array}
$$

Geometry

The **perimeter** of a polygon is the distance around the outside of the figure. To find the perimeter, add the lengths of the sides of the figure. Be sure to label your answer.

Perimeter = sum of the sides

Example: Find the perimeter of the rectangle below.

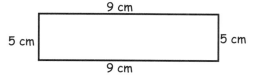

Perimeter = 5 cm + 9 cm + 5 cm + 9 cm
Perimeter = 28 cm

Example: Find the perimeter of the regular pentagon below.

A pentagon has 5 sides. Each of the sides is 4 m long.

P = 4 m + 4 m + 4 m + 4 m + 4 m
P = 5 × 4 m
P = 20 m

Help Pages

Who Knows???

Sides in a Quadrilateral? .... (4)

Sides in a Pentagon? .... (5)

Sides in a Hexagon? .... (6)

Sides in an Octagon? .... (8)

Inches in a foot? .... (12)

Feet in a yard? .... (3)

Inches in a yard? .... (36)

Ounces in a pound? .... (16)

Pounds in a ton? .... (2000)

Cups in a pint? .... (2)

Pints in a quart? .... (2)

Quarts in a gallon? .... (4)

Years in a decade? .... (10)

Figures with the same size and shape? ....
.... (congruent)

Figures with same shape, but different size? ....
.... (similar)

Answer to an addition problem? .... (sum)

Answer to a subtraction problem? .... (difference)

Answer to a multiplication problem? ...... (product)

Answer to a division problem? .... (quotient)

TPS 121185